A B C
intelligence

complete with

| + |

 =

.....

complete with

$$2 + 2$$

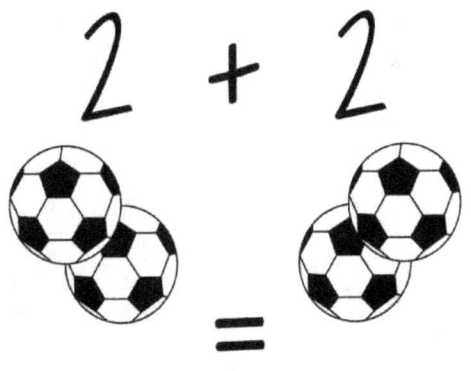

$$=$$

.....

complete with

$$3 + 3$$

=

.....

complete with

color balloon

color balloon

complete with

color balloon

complete with

color balloon

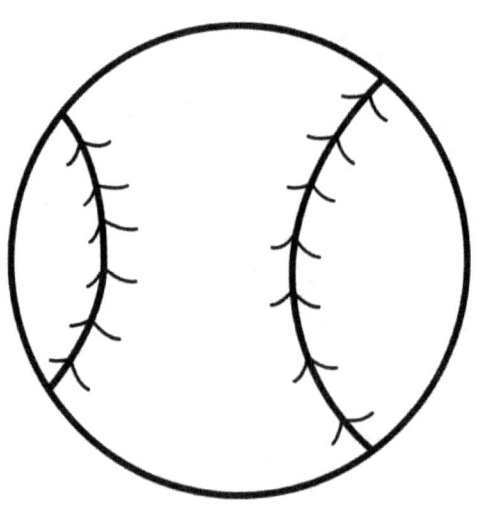

complete with

color cat

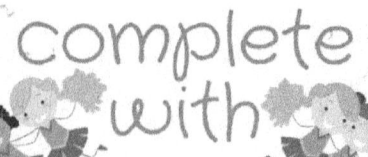

complete with

color Pumpkin

complete with

color Pumpkin

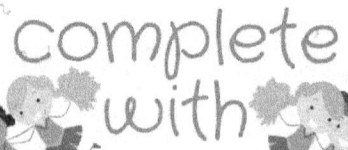

complete with

color Pumpkin

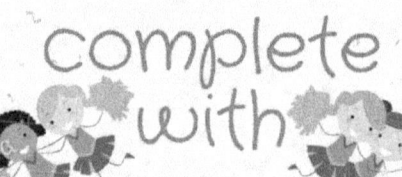

complete
with

complete
with

color

complete

with

student

complete with

color

complete
with

..... + ...

=
.....

complete
with

complete
with

color

complete with

..... + ...

=
.....

complete with

..... + ...

=
.....

complete
with

color

complete
with

..... + ...

=

.....

complete with

..... + ...

=

.....

complete with

$$..... + ... $$

$$= $$

$$..... $$

complete
with

complete with

color

complete
with

..... + ...

=
....

complete
with

..... + ...

=

.....

complete
with

color

complete
with

..... + ...

=

.....

complete with

..... + ...

=
.....

complete with

..... + ...

=

.....

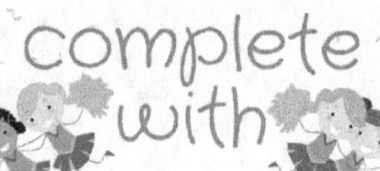

complete with

complete
with

color

color balloon

complete with

color balloon

complete
with

complete
with

color

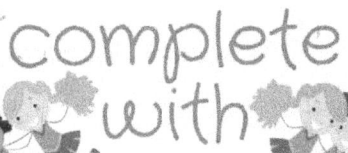

complete with

color balloon

complete with

color balloon

complete with

color balloon

complete
with

color

complete with

color cat

complete
with

color

complete
with

color cat

complete with

color cat

complete with

complete
with

color

complete with

color cat

complete with

color cat

complete with

color Pumpkin

complete with

color Pumpkin

complete
with

color

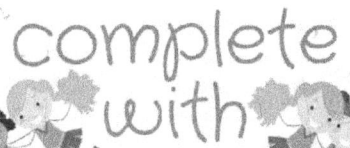

complete with

color Pumpkin

complete with

color Pumpkin

complete with

color Pumpkin

color Pumpkin

complete
with

complete
with

color

complete with

color Pumpkin

complete with

color Pumpkin

complete
with

color Pumpkin

complete
with

complete with

color

complete with

color Pumpkin

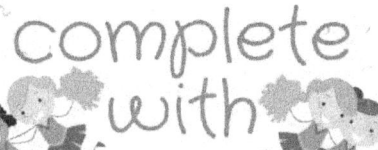

complete with

color Pumpkin

complete
with

color Pumpkin

complete with

complete
with

color

complete with

color Pumpkin

complete with

color Pumpkin

complete with

color Pumpkin

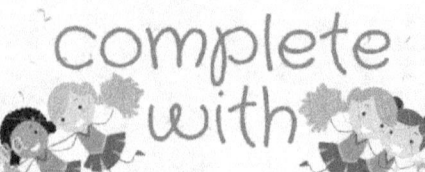

complete
with

color

complete with

color Pumpkin

complete with

color Pumpkin

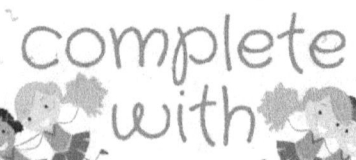

complete with

color Pumpkin

complete
with

color

color balloon

complete with

color balloon

complete with

color balloon

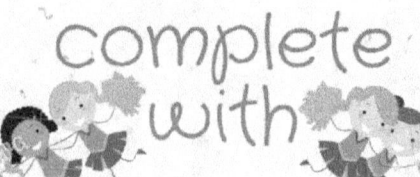

complete with

complete
with

color

color balloon

complete with

color balloon

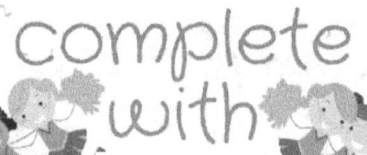

complete with

color balloon

complete with

color balloon

complete with

complete with

color

complete with

color balloon

color balloon

color balloon

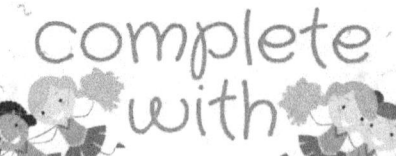

color balloon

complete with

color balloon

complete with

complete
with

color

A B C
intelligence